I'm TWO Times Tabler!

Bill Gillham and Mark Burgess

A Magnet Book

Where are all these pigs going?

Why are they all in twos ?

I'm a two times tabler! Are you?
I'm a two times tabler
2X

$1 \times 2 = 2$

Polite pigs don't put their trotters on the table.
Not pig swill again!
PIG NEWS

$2 \times 2 = 4$

two twos are four

She says my bedroom's a pigsty.
He's squealing on me!
PIG NEWS

$3 \times 2 = 6$

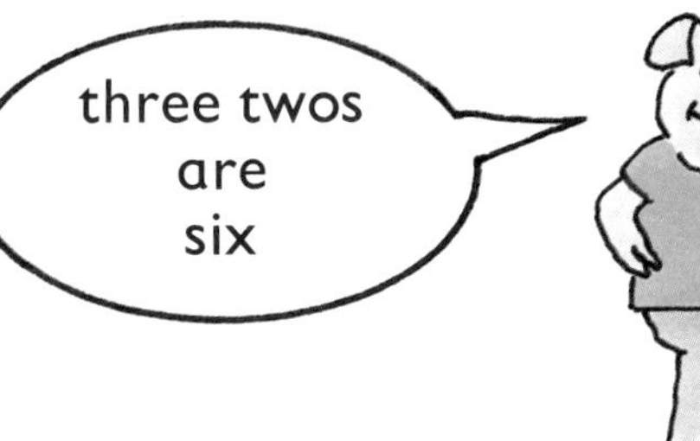
three twos are six

You're making a perfect pig of yourself!
Little pigs should be seen and not heard.
PIG NEWS

$4 \times 2 = 8$

four twos are eight

We're expecting a litter very soon.
Ssh— little pigs have big ears...
PIG NEWS

$5 \times 2 = 10$

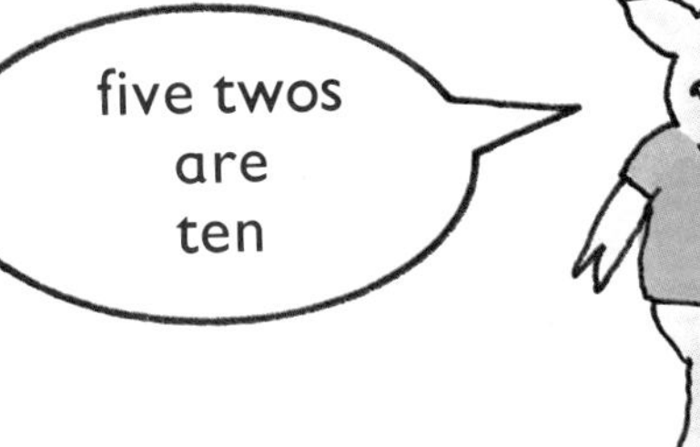
five twos are ten

This little piggy went to market!
This little piggy had roast beef!

$6 \times 2 = 12$

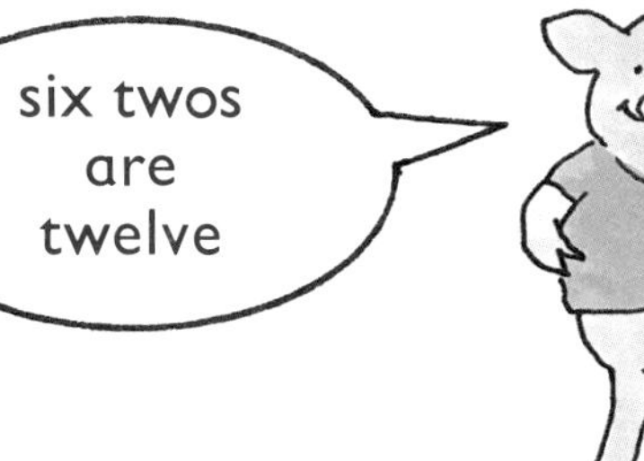
six twos are twelve

I'm trying to keep my figure.
I'd rather have acorns!

$7 \times 2 = 14$

seven twos are fourteen

Piglets aren't what they used to be...
I need to know my tables with all the piglets I've had !

$8 \times 2 = 16$

eight twos are sixteen

I married the Owl and the Pussycat last week !
Here's something for your piggy bank !

$$9 \times 2 = 18$$

nine twos are eighteen

He's getting very pig-headed!
I'm piggy in the middle! Ha! Ha!

10 × 2 = 20

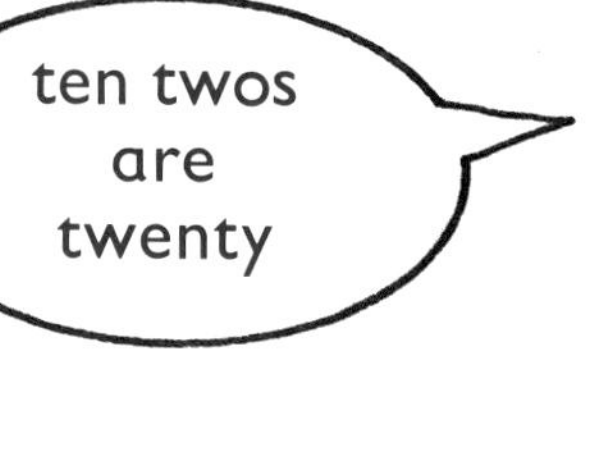
ten twos are twenty

You're my sugar pig!
He's just an old ham!

$11 \times 2 = 22$

eleven twos are twenty-two

There was a pig went out to dig!
We're just one big happy family!

$$12 \times 2 = 24$$

twelve twos are twenty-four

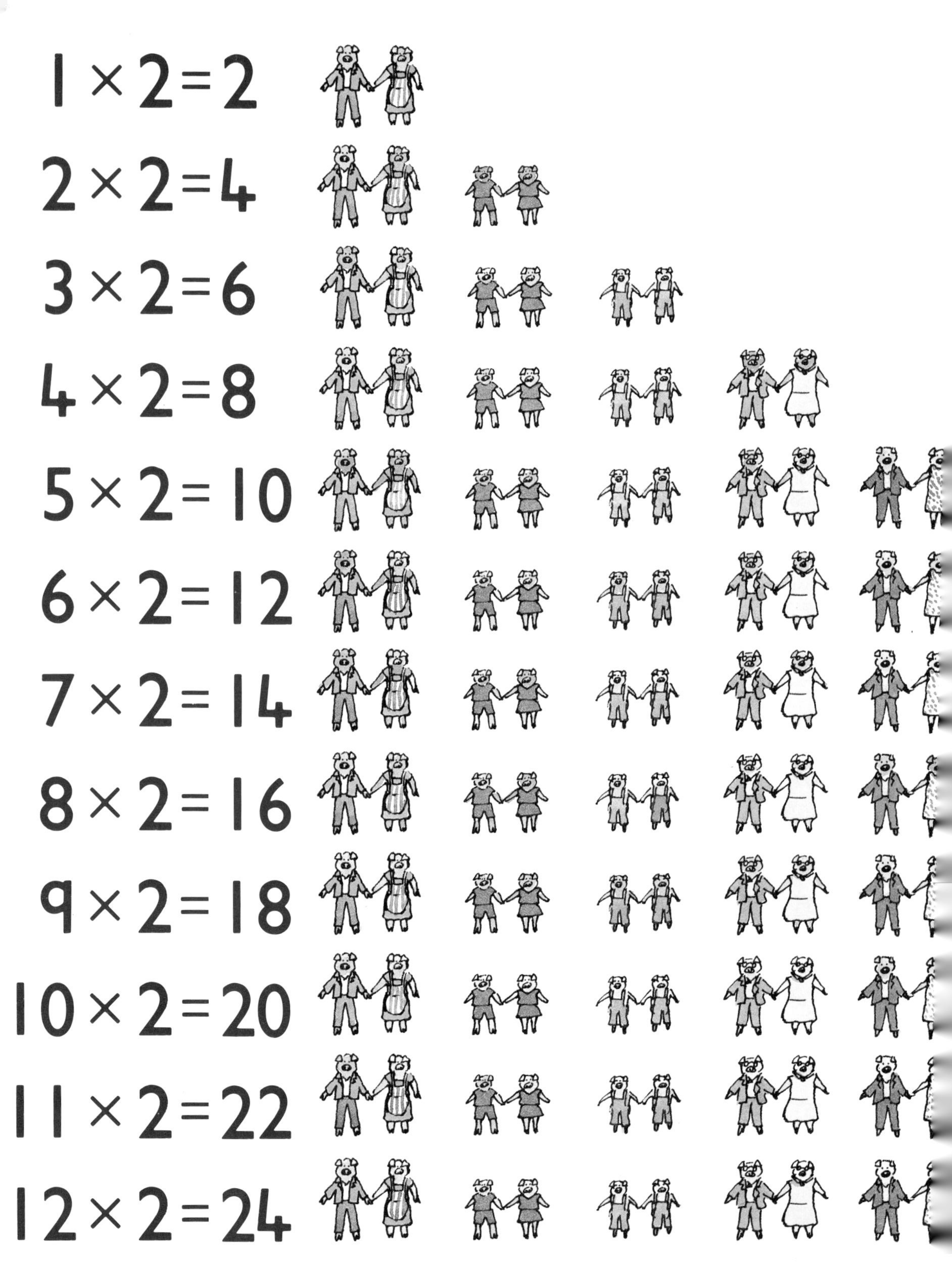

$1 \times 2 = 2$

$2 \times 2 = 4$

$3 \times 2 = 6$

$4 \times 2 = 8$

$5 \times 2 = 10$

$6 \times 2 = 12$

$7 \times 2 = 14$

$8 \times 2 = 16$

$9 \times 2 = 18$

$10 \times 2 = 20$

$11 \times 2 = 22$

$12 \times 2 = 24$

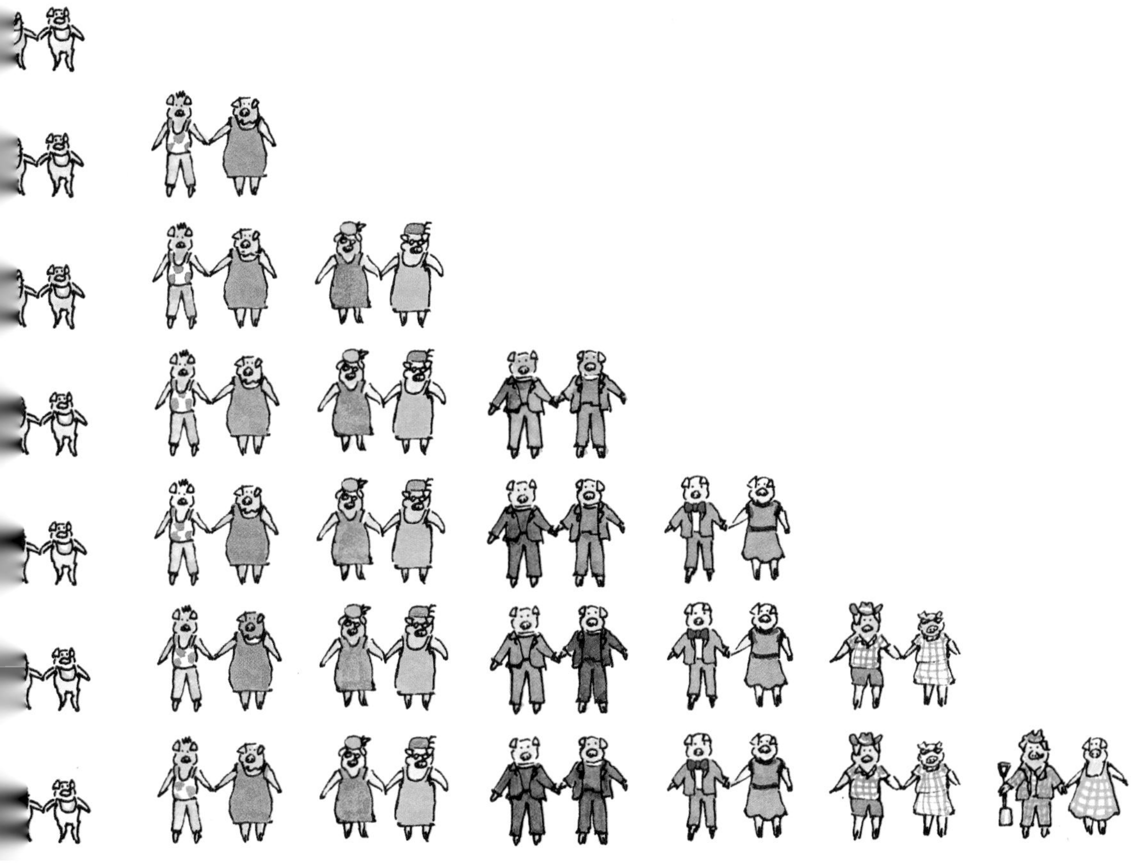

Activities

TWO TIMES TABLE PYRAMID

Using Smarties, or coloured plastic counters, make a number pyramid, saying the two times table as you go:

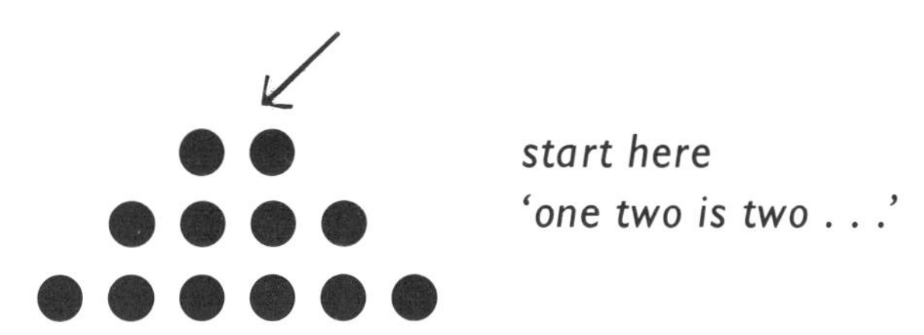

Let the child select the sweets, picking two of the same colour for each pair, and making each pair a different colour from the pairs immediately next to it.

ROW-OF-HOUSES NUMBER LINE

Make a row of twelve houses by folding and cutting a strip of paper. Write *large* even numbers 2 to 24 on the houses, and stand them up concertina fashion. Details like doors and windows can be added later.

Explain that when you say the two times table you are *counting* in twos: 2 – 4 – 6 etc.
Practise this and make another set of houses for the child to put in the numbers.
You can reinforce the activity when walking along an appropriately numbered street.

NUMBER SETS

With the child helping, count out 24 beads (beans, counters, sweets) and find 12 little compartments, such as two egg-boxes or cake tins.

Show the child how to count two objects
into each individual compartment,
saying the two times table as he does so.

Then ask, 'How many twos have we got altogether?'
or, 'How many twos in this egg-box?'

NATURAL PAIRS

Get together twelve pairs of shoes, boots, and slippers (all different sizes) and muddle them up in a heap.

Ask the child to pick out the matching pairs and set them out in a line. He can then walk along the line saying the two times table.

Find a large cardboard box for him to put them in, this time *counting* in twos.

Children need to know their tables because:
– simple multiplication, *which you can do in your head,* is a skill of practical use in everyday life;
– the number patterns and groupings that occur in tables help them to understand more advanced mathematical concepts like *sets*, number *series* and *progressions*.

The Times Table Books teach these ideas in a clear and enjoyable fashion and show vividly what happens when you multiply.

Dr Bill Gillham is senior lecturer in the Department of Psychology at Strathclyde University.

First published in Great Britain in 1987
as a Magnet original
by Methuen Children's Books Ltd
11 New Fetter Lane, London EC4P 4EE

Printed in Great Britain

ISBN 0 416 00202 1